Complex research of environm

Shafaq Baghirova

Complex research of environmental problems in the Absheron peninsula

LAP LAMBERT Academic Publishing

Cover image: www.ingimage.com

Publisher:
LAP LAMBERT Academic Publishing
is a trademark of
International Book Market Service Ltd., member of OmniScriptum Publishing Group
17 Meldrum Street, Beau Bassin 71504, Mauritius

Printed at: see last page
ISBN: 978-620-0-54884-9

Complex researches of environmental problems in the Absheron peninsula

Content

Introduction

"Since the rapid change of the biosphere as a result of anthropogenic impact is a global catastrophe, the anthropogenic landscape can be viewed as the landscape of the future. It is possible that after 100-200 years, the anthropogenic landscape will cover all areas of the surface, except for permanent congestion and high mountainous areas. This could be attributed to the high natural growth of the world population, the unstable development of industry and agriculture, the increased demand for energy resources, and so on. As a result, humankind will be faced with the creation of new sustainable landscapes that can provide it with oxygen, water, food and energy. Without the protection of the gene pool of animals and plants, without knowing the basic laws of nature that are the basis of the ecosystem and biodiversity in general, we will be helpless and helpless."

(Flint V.E., 2004)

In recent years, as a result of the rapid development of scientific and technological progress and active activities of people, a number of harmful substances in the environment have been released into the environment by polluting the biosphere. Such processes undermine the ecological balance of the area and negatively affect the normal activities of people.

It is known that the Absheron peninsula, including Baku, is located in the semi-desert zone. The area of the Absheron peninsula is mainly ephemeral. Trees are almost too few. The natural flora of Absheron has 8 family, 22 species of trees and shrubs of 13 genus. Absheron's existing natural vegetation, which is contaminated with environment, does not play a significant role in cleaning the atmosphere and regulating the region's ecological balance.

Therefore, it is important to take advantage of new plant species with a wide range of decorative tolerance instead of research in environmental regulation. Between indicators are per capita greenery, which defines quality of life of the urban population. In Paris this figure is 6 square meters / person, in New York - 7.5 square meters / person, in Moscow - 20 square meters / person.(Melexova O.P.,2002). As you know, compared to developed and developing countries, per capita greenery in Azerbaijan and in Absheron is smaller. Therefore, enriching the diversity of plants on the Absheron peninsula is one of the most pressing issues facing researchers and other organizations in the modern era. In this regard, the focus is on enriching the peninsula and the city of Baku with new species and determining the viability of these species. After Azerbaijan gained its independence, extensive landscaping activities have been carried out with the use of new plant species to enhance and protect biodiversity and preserve the ecological balance. New parks and gardens are being built in Baku and surrounding areas, using various new trees and shrubs. By 1985, Baku's greening used 278 species of trees and shrubs, with 153 species of trees and 125 species of shrubs. Currently, the number of firewood plants used is between 800-1000. However, the scientific basis, biological features, ornamentation, durability and long-term viability of new trees and shrubs imported from many foreign countries in the dry soil climate of the Absheron have not been fully studied. As a result, under certain conditions, certain plants are destroyed.

Construction of Baku on the basis of landscape architecture is one of the most pressing issues of the present. It is known that the Absheron peninsula is rich in oil and chemical industry. The area is not rich in taxonomic composition of natural trees and shrubs. Newly introduced wood and shrubs from this pattern play an important role in the formation of the Absheron flora. The use of many ornamental plants in greening around the cities and towns of the Absheron peninsula, the newly built industry, residential areas, roads and bridges, along with decorative architecture, is also a key factor in improving social living and recreation. This type of plant also plays an important role in the protection of the plant gene pool, in the enhancement of biodiversity and in the restoration of ecological balance. It is important to note that plants are also a source of aesthetic effects for humans.This type of greenery is particularly important in the severe psychological conditions of cities and in the protection of modern environmental pollutants. Many introductions have reclamation significance along with these features.

They play a major role in preventing sliding slopes around the Caspian Sea, water erosion in the soil, strengthening of sandy areas.

Actions taken by individuals and the government as an important step towards the protection of life support systems, along with the wellbeing of people, also provide a rich variety of life on our planet.

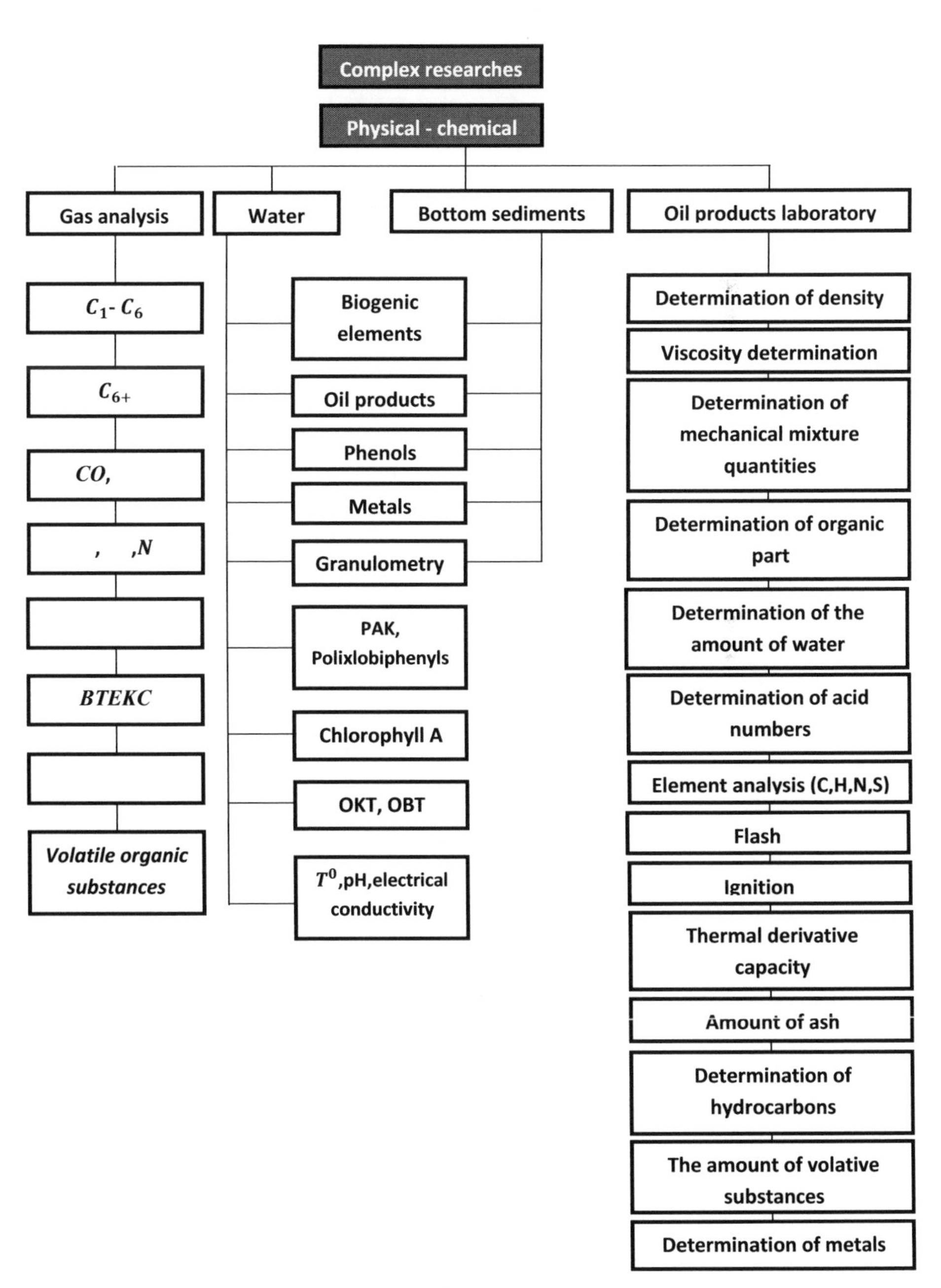
Complex researches
Physical - chemical
Gas analysis
Water
Bottom sediments
Oil products laboratory
C_1- C_6
C_{6+}
CO,
, ,N
BTEKC
Volatile organic substances
Biogenic elements
Oil products
Phenols
Metals
Granulometry
PAK, Polixlobiphenyls
Chlorophyll A
OKT, OBT
T^0,pH,electrical conductivity
Determination of density
Viscosity determination
Determination of mechanical mixture quantities
Determination of organic part
Determination of the amount of water
Determination of acid numbers
Element analysis (C,H,N,S)
Flash
Ignition
Thermal derivative capacity
Amount of ash
Determination of hydrocarbons
The amount of volative substances
Determination of metals

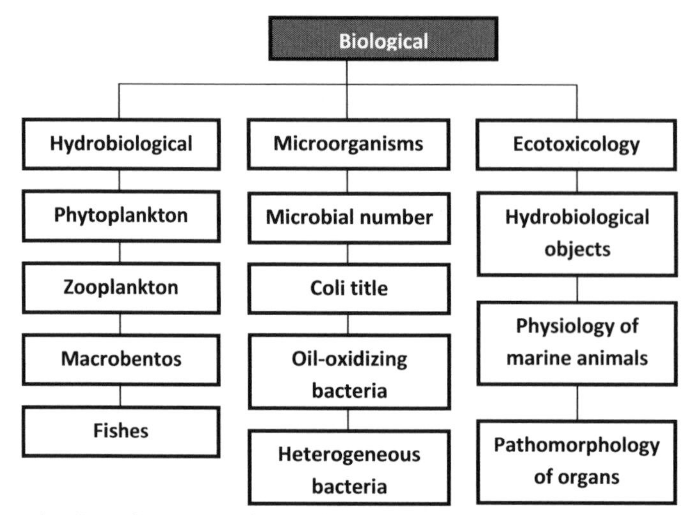

The Complex Researches Laboratory (CRL) was established in 1998 on the basis of the Complex Research Laboratory of the SOCAR Environmental Research Expedition of the Complex Geological Exploration and Topographycal Department. The CRL has passed accreditation in the Azerbaijan National Certification System laboratory carrying out onshore and offshore environmental studies. The Complex Researches Laboratory was established and equipped with the purpose of rendering wide spectrum analytical services for offshore and onshore environmental studies. The CRL specialized in collection and analysis of samples for the following types of studies, carried out both onshore and offshore:

- Environmental Baseline Study (EBS)
- Environmental Impact Assessment (EIA)
- Ecotoxicological Researches of Caspian organisms

The CRL is placed in a new building and equipped with modern laboratory facilities and computers. The CRL personnel comprise highly qualified specialists and scientists, who have a significant work experience and young specialists developing their knowledge continually. These scientists participated in the majority of ecological projects carried out for the various foreign operating oil companies in the Azerbaijan sector of the Caspian Sea. Personnel of the CRL consists of chemists, biologists-taxonomists, microbiologists and toxicologists. Technical trainings on modern equipment provided to CRL employees continually improves their knowledge and skills. The CRL uses international methods and standards while carrying out wide spectrum of environmental analysis. The CRL uses a specially equipped offshore vessel "M.Suleymanov" "MPK-452" for its offshore researches, this vessel presents a mobile laboratory specially equipped with devices and computers for collecting offshore sediments samples up to the depth of 1000 meters. The Complex Research Laboratory renders the following services:

- Environmental offshore monitorings
- Analytical studies of water (sea, surface, produced water, waste water) and ground
- Ecological assessment of biota communities
- Analysis of taxonomy of periphyton, macrozoobenthos, phytoplankton, zooplankton and ichtyofauna
- Ecotoxicological testing of animal tissues
- Analysis of microorganisms taxonomy and study of biological decomposition
- Chemical analysis of water, soil, offshore sediments, animal tissue and plants
- Provision of information about condition of Caspian Sea biological objects and other various environmental parameters of the Caspian Sea
- Laboratory services on the scientific research vessel

The Complex Research Laboratory is furnished with modern equipment and facilities. Analysis orders of taken samples for investigation of the current environmental condition and environmental impact assessment on production sites of SOCAR enterprises, and also other international and local companies are being performed according to international standards in the laboratory.
Complex Research Laboratory was accredited on international standard TS EN ISO/IEC 17025:2012 certifying the accuracy of the analysis and awarded certificate AB-0550-T.
The lab analyzes the samples of BP,TOTAL E&P Absheron B.V. and other oil and gas companies appropriate to the international standards.

Ecological monitoring activities

Identify an environmental preliminary situation in the exploration and processing of onshore and offshore oil and gas fields, environmental impact assessment (EIA) and multidisciplinary environmental monitoring are carried out by Ecology Department.

Monitoring is focused on the study, assessment of the environmental situation in onshore and offshore oil and gas fields, identify the impacts of production into the ecosystem, and submitting the proposals to reduce current ecological problems.

Monitoring is carried out by the Expedition Team consisting of the specialized abd qualified experts. The expedition team is also provided with modern facilities and equipment. Information about results of monitoring is elucidated by mass media.

Onshore oil and gas monitoring are focused on the followings:

- Detailed study of areas polluted with the waste generated in the process of oil, oil products
- Monitoring of produced and domestic-faecal waters
- Control over the condition of treating plants and equipment
- Control over atmosphere air, soil and quality indicators of water, research of its physical-chemical parameters
- Observing the environmental changes in current operations
- Quantitative and compositional study of pollution substances
- Detection the sources of pollution
- Preparation of proposals for the elimination of pollution

Identify the impacts of oil and gas production to Caspian ecosystem, determine the physical and chemical parameters of water environment, lithology and grain-size composition of bottom sediments, the taxonomy of microorganisms are examined and anthropogenic impact is also studied.

Offshore monitoring is carried out twice a year: in spring and autumn.

The followings are studied to identify the impacts to Caspian ecosystem:

- Physical-chemical parameters of the water
- Biogenic elements and oil products
- Physical-chemical analyses of bottom sediments
- Grain-size composition of bottom sediments
- Organic substances and metal accumulation
- Taxonomy of microorganisms
- Development dynamics of the flood chain of the microorganisms in the aquatic environment

Marine vessels of offshore monitoring studies:

- M. Suleymanov
- MPK-452
- "Adventure" type boat

MPK-452 vessel

M.Suleymanov vessel

“ADVENTURE” type boat

Preparation of ESIA documents

Ecology Department prepares the document of the Environmental and Social Impact Assessment (ESIA) according to the projects implemented by SOCAR's enterprises. The document is approved after the submission to the public discussion, is complied accoding to the projects implemented by SOCAR's enterprises.
The document is approved after the submission to the public discussion, is compiled according to the current legislative of Azerbaijan Republic.

Recent projects in this field are the followings:

- ESIA documents over construction of Stationary Deep Water Jackets, drilling and exploitation of wells located onshore and offshore oil and gas production facilities of Azneft PU
- ESIA document over construction and exploitation of Carbamide Plant
- ESIA document over expansion of Waste Management Center
- ESIA document over modernization and exploitation of EP-300 plant of Azerkimya PU
- ESIA document document over modernization and reconstruction project of the Heydar Aliyev Oil Refinery
- Environmental Baseline Survey of Absheron Field Development for TOTAL E&P Absheron B.V.
- ESIA for Garabakh field development

Preparation of Ecological normative documents

SOCAR has a corporate "environmental policy" document, which defines areas of partical activities in the sphere of environment, ecological safety, and efficient use of natural resources; the principle of zero emissions was declared.
Development of Ecological normative documents is implemented according to the Law of 'Regarding the Environmental Protection' dated 2000 and the Resolution of the Cabinet of Ministers of Azerbaijan Republic on the approval of normative and legal documents on the permissible limits of the thickness of harmful substances in the atmospheric air, dated 08.06.1999.

Ecological normative document's preparation group operate at the Ecology department that is specialized in this sector. The team is consisting of qualified experts and supplied with the special software. The group prepares the following ecological normative documents for SOCAR's enterprises and organizations:

- Ecological Passport
- Passport of Hazardous Waste
- Limit on the Permissible Waste
- Limit on the Permissible Discharges

Ecological Database

The new operative electronic database has been created for the acquired information in the frame of monitoring activities carried out on the activity sites of the oil and gas extraction departments of the Absheron Peninsula.

Extensive ecological monitoring activities have been carried out, various information sources and laboratory studies have been used during preparation of the ecological database. Thus, land areas contaminated with oil and other waste, areas remaining under produced waters on the Absheron Peninsula, and their depths of contamination is reflected as per exact coordinates in the ecological database. The "Arc View" program provides an opportunity for placing the results of analysis on coordinates of carried out ecological monitoring activities on the map and also to continuously renew acquired results during repeated monitoring activities.

According to the ecological database, it is possible to observe and assess the change tendency of the ecological condition, detection of pollution sources and also manage environmental issues in the areas, where monitoring has been carried out.

2010 in Azerbaijan was announced as year of ecology, and numerous programs for the implementation of environmental projects encompassing the whole country were taken. From this date, implementation of megaprojects in the sphere of environment began in the Azerbaijan Republic, and management of the environment became one of the priorities of the state's sustainable development strategy.

Recently, SOCAR carried out a physical inventory of lands contaminated with oil and oil products during the 165-year history of the Azerbaijan oil industry.

Inventory was carried out accounting for results of detailed field survey both onshore and offshore. Contaminated areas were shown on the map, with the indication of nature of pollution, penetration depth, and geography.

Ecological database for contaminated lands was developed using special soft- ware program ArcGIS.

This database is annually updated based on completed work on rehabilitation and recultivation, as well as results of the environmental monitoring

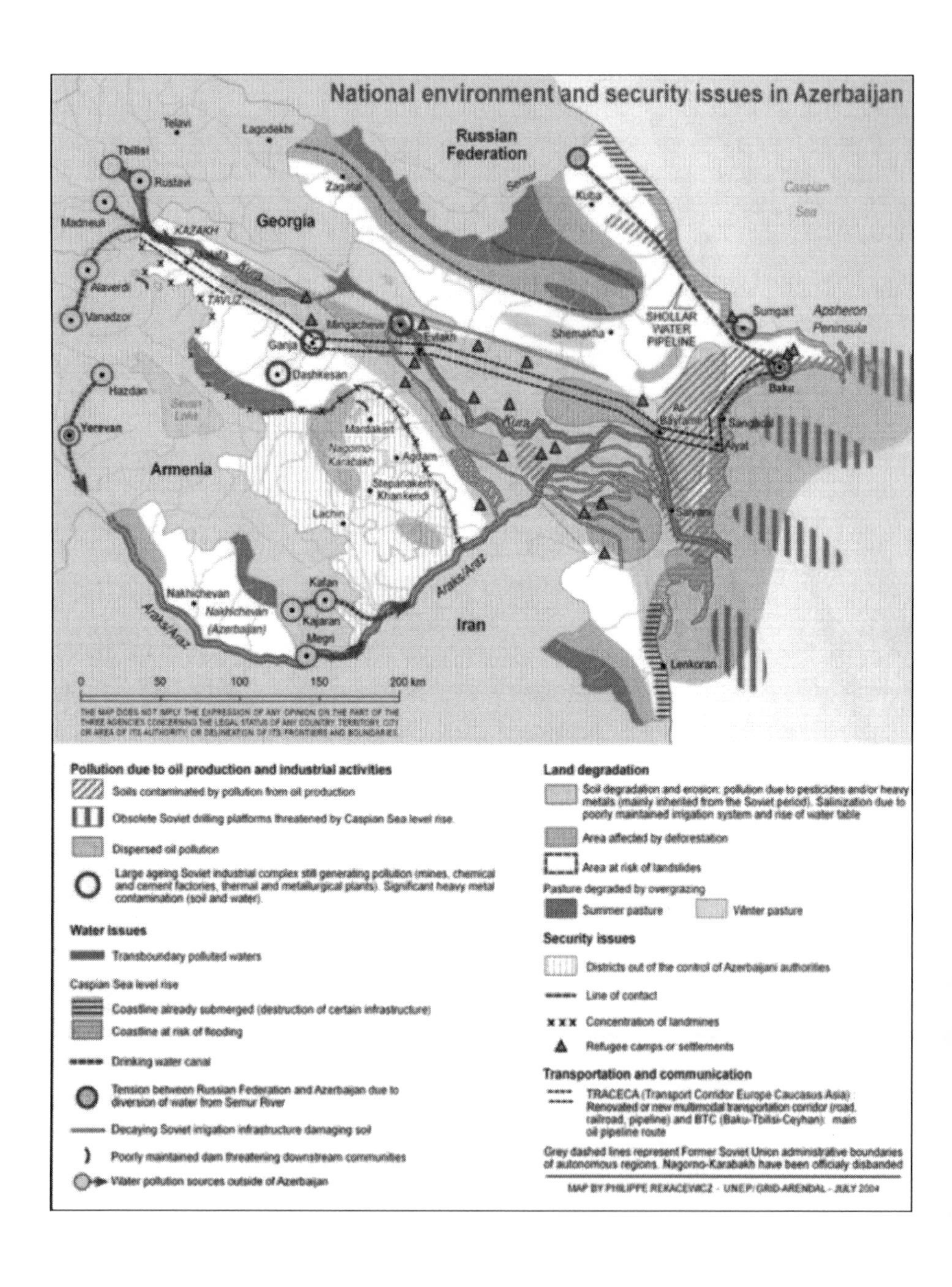
National environment and security issues in Azerbaijan
Russian Federation
Georgia
Armenia
Iran
Caspian Sea
Apsheron Peninsula
Telavi
Lagodekhi
Tbilisi
Rustavi
Madneuli
Alaverdi
Vanadzor
Hazdan
Yerevan
Zagatal
Kuba
Samur
Sumgait
Baku
Shemakha
SHOLLAR WATER PIPELINE
Mingachevir
Yevlakh
Ganja
Dashkesan
KAZAKH
TAVUZ
Kura
Ali-Bayramli
Sangachal
Alyat
Mardakert
Nagorno-Karabakh
Agdam
Stepanakert Khankendi
Lachin
Salyan
Lenkoran
Nakhichevan
Nakhichevan (Azerbaijan)
Kafan
Kajaran
Megri
Araks/Araz
0 50 100 150 200 km
Pollution due to oil production and industrial activities
Soils contaminated by pollution from oil production
Obsolete Soviet drilling platforms threatened by Caspian Sea level rise
Dispersed oil pollution
Large ageing Soviet industrial complex still generating pollution (mines, chemical and cement factories, thermal and metallurgical plants). Significant heavy metal contamination (soil and water).
Water issues
Transboundary polluted waters
Caspian Sea level rise
Coastline already submerged (destruction of certain infrastructure)
Coastline at risk of flooding
Drinking water canal
Tension between Russian Federation and Azerbaijan due to diversion of water from Semur River
Decaying Soviet irrigation infrastructure damaging soil
Poorly maintained dam threatening downstream communities
Water pollution sources outside of Azerbaijan
Land degradation
Soil degradation and erosion: pollution due to pesticides and/or heavy metals (mainly inherited from the Soviet period). Salinization due to poorly maintained irrigation system and rise of water table
Area affected by deforestation
Area at risk of landslides
Pasture degraded by overgrazing
Summer pasture
Winter pasture
Security issues
Districts out of the control of Azerbaijani authorities
Line of contact
Concentration of landmines
Refugee camps or settlements
Transportation and communication
TRACECA (Transport Corridor Europe Caucasus Asia): Renovated or new multimodal transportation corridor (road, railroad, pipeline) and BTC (Baku-Tbilisi-Ceyhan): main oil pipeline route
Grey dashed lines represent Former Soviet Union administrative boundaries of autonomous regions. Nagorno-Karabakh have been officialy disbanded
MAP BY PHILIPPE REKACEWICZ - UNEP/GRID-ARENDAL - JULY 2004

Waste Management

The Department of Ecology of the State Oil Company of the Republic of Azerbaijan is responsible for environmental protection, environmental safety and natural resources in accordance with the "Comprehensive Action Plan for the Improvement of the Ecological Situation in the Republic of Azerbaijan for 2006-2010" approved by the Presidential Decree of September 28, 2006. defined general directions of practical activities in the field of use. In addition, the President of the State Oil Company of the Republic of Azerbaijan has approved the Environmental Policy. The document reflects the adherence to the principle of "zero waste" by assessing the initial state of the environment in new areas of activity and minimizing its negative impacts.

One of the main directions activities on the provision of environmental protection and ecological safety is the waste management in accordance with the requirements of modern standards. The waste Management Center (WMC) Expansion Project of the Ecology Department located in the Garadagh district was implemented toward to the acceptance, treatment and comprehensive utilization of waste formed in SOCAR's activity areas.

As a result of the project executive the WMC has been equipped with the most modern unit, for the treatment of drilling cuttings in accordance with International Standards, 2 "VacuDry" units manufactured in Germany have been installed. Additionally, 4 pits 50000 m3 capacity were constructed for disposing of waste formed as a result of drill cuttings treatment.

The most modern environmental standards of Azerbaijan, European Union and USA have been applied for construction and exploitation of WMC.

This project provides the waste management by covering the whole process such as registration, sorting, collection, treatment and disposal of waste formed in oil and gas, petrochemical industry.

The WMC inaugurated with the was participation of representatives of MENR, SOCAR and BP, as well as other companies operating in the oil and gas sector in September 2016.

The WMC is the first and only object in CIS and Caspian region on management and treatment of waste formed in the oil industry.

The treatment and disposal drill cuttings in "VacuDry" units, returning drilling reagents to recycling by recovering drilling reagents, deposition of residual waste in the appropriate pit are implemented in WMC.

In addition, different types of solid industrial waste formed in departments and enterprises of SOCAR is taken to WMC and stored in appropriate pits for recycling in the future. Also, control monitoring is carried out and the reports are prepared and submitted to the relevant authorities by the experts of Ecology Department for the waste management in order the relevant legislation, additionally, according Management Plan.

Atmosphere emission reduction.

As in other countries, environmental protection and natural resources issues in the Republic of Azerbaijan have come to the fore in recent years. The main sources of CO2 emissions in the Republic of Azerbaijan are the energy and industrial sectors. While in the energy sector CO2 emissions come from the burning of fuel, including in the production of energy, oil and gas extraction, transport, and human settlements, the largest source of CO2 emissions in the industrial sector have been from mineral production and metallurgy.

Methane (CH4) is emitted by nearly all sectors of the Azerbaijani economy but has increased mostly in the agricultural sector due to fermentation and manure, and in the waste sector due to human settlements. Nitrous oxide (N20) emissions in Azerbaijan have declined since 1990; in 2005 emissions were reduced by 64% from 992 Gt CO2 equivalents in 1990.

Emissions of halogen substances per-fluorocarbon, hydro-fluorocarbon and sulfur hexafluoride are not found at significant levels in Azerbaijan.

The Republic of Azerbaijan is a non-Annex I member of the Convention and, therefore, does not have any binding obligations to reduce emissions of greenhouse gas (GHG). No concrete policies or laws on carbon market issues have been established to date to implement the Kyoto Protocol. However, Law No. 109-I1Q of the Republic of Azerbaijan On Protection of the Atmosphere, dated March 27, 2001, requires that a special emissions permit be obtained from the Ministry of Ecology and Natural Resources of the Republic of Azerbaijan (MENR) prior to the emission of hazardous materials into the atmosphere. Such an emissions permit is the only emission allowance in Azerbaijan. As a matter of law, permits may be suspended or cancelled in case of violations; a permit is not transferable.

To achieve the purposes of the Kyoto Protocol, the Republic of Azerbaijan has entered into memorandums of understanding with the Federal Republic of Germany and the Kingdom of Denmark on expanding vegetative cover and on capacity-building towards GHG emissions reduction.

In accordance with the instruction of the Azerbaijan Republic President of Azerbaijan Republic expediting transition to the Euro-standards, the vehicles of SOCAR's departments and enterprises pass through technical inspection at 2 ecological measurement stations created in the Sabunchu and Surakhany districts of Baku city with the purpose of the measuring the hazardous gases vented from the vehicles of

SOCAR's department and enterprises, the intensification of the inspection over the vehicles' technical condition and ensuring the correspondence of the emission with the standards.

The vehicles belonging to SOCAR regularly pass through the aforementioned inspection and special emblems are attached onto the vehicles that meet ecological requirements.

Reduction of the impacts on Climate Change.

In Azerbaijan, the leading source of carbon dioxide, the heat-trapping greenhouse gas (GHG) that has severely driven recent global warming is associated with the oil and gas sector, the main driver of national economy. Carbon emissions occur in the petroleum production and end-use energy consumption sectors, due to the combustion of fossil fuel during the oil and natural gass extraction processes, transportation, and in residential facilities, which, altogether have an overall negative impact on global warming and climate change.
GHG emissions reduction activities are implemented through application of less waste-making technologies, detection of carbon intensive manufacturing process, energy effective measures and planting which is in line with approved in 2010 " Climate Change mitigation strategy of SOCAR" for the period 2010-2020.

To address this environmental challenge, since 2015 UNDP and the State Oil Company of Azerbaijan (SOCAR) have been implementing a series of measures to reduce the amount of GHG released into the atmosphere.

The energy efficiency measures are part of the Nationally Appropriate Mitigation Actions project (NAMA) for low-carbon end-use sectors in Azerbaijan. Funded by the Global Environment Facility (GEF), the project is to continue through 2020 and will draw upon three key areas of intervention. These interventions include the application of energy-efficient technologies in construction design and architecture, enforcement of sustainable driving habits and the provision of clean sources of fuel for cooking and other household purposes.
Through joint effors, the total area of 10,000 square meters in SOCAR administrative and service buildings has been appointed with energy-saving equipment and environmentally friendly appliances, such as energy-efficent windows, floor heating, energy-efficent ventilation and heat insulation systems, to name a few.Each year, these buildings will benefit approximately 2,500 people.The Green buildings project carried out repair work using energy efficient technologies in the administrative and social buildings of SOCAR. As a result, energy consumption decreased by 30-40%.
In addition to green building design, safe snd energy-efficient driving technologies were introduced in Baku, including eco-driving simulators and hybrid cars, which help transform energy from fuel combustion into electric energy.Coupled with a series of specialised training for approximately 1,200 drivers,these sustainable transport performance measures are estimated to result in the 10 to 15 percent of reduction in carbon dioxide emission from vehicles.

In 2017, SOCAR have approved "Associated gas reduction Plan at SOCAR and in projects in which SOCAR takes part". The newly- developed Plan covers the period 2017-2022. a result of measures proposed by structural divisions of SOCAR, Operating companies and Joint ventures, up to 300 million cubic meters of associated gas is expected to be gathered per year beginning from 2022.

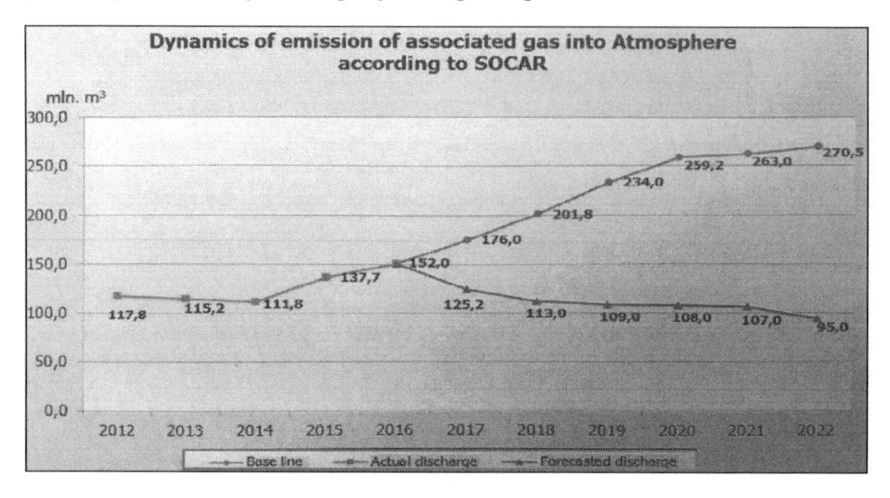

Renewable energy sources

The Republic of Azerbaijan is, by its geographical location and climate conditions, one of those countries that have favorable conditions to use sources of alternative energy.

Various parts of the country have good quality resources for renewable energy. The Absheron peninsula, the coastal areas along the Caspian Sea, as well as the Caspian basin, the Nakhchivan Autonomous Republic and Baku are considered to be suitable for developing wind energy projects. According to the experts analysis wind in the Azerbaijan blows more than 250 out of 365 day per year. It is also best profitable due to the cost, ecological cleanness and its renewable properties compared to other alternative energy sources.

Baku, Kura-Araz lowland, Absheron Peninsula and the Nakhchivan Autonomous Republic are considered to be suitable for developing solar power projects. According to investigation made by experts there are 300 sunny days in Azerbaijan. The quantity of Sun lights are more than all reserves of crude oil, gas and cool. The effectiveness and advantage of solar energy is no CO_2 emission into atmosphere.
Total potential capacity of solar power is estimated at 8,000MW, wind-15,000MW, biomass-900MW,geothermal-800MW, and small hydro-600MV.
The first priority in moving towards a sustainable energy system is to increase the efficiency of energy use.This consists in obtaining tha same result in terms of energy services with less energy. Efficiency can be improved and energy can thus be saved in all types of energy utilisation, and, since energy is used in all human activities, essentially one can save energy in all human activities.
The resolution of the President of the Republic of Azerbaijan I.Aliyev on "The State Program for Use of Alternative and Renewable Energy Sources in the Republic of Azerbaijan" proves that in this issue Azerbaijan has expressed its solidarity with the world community.

Despite ranking in the top 20 among nations with crude oil proved reserves and the top 25 when it comes to natural gas, Azerbaijan is moving ahead with developing renewable sources of energy.

Although SOCAR have been producing fossil fules, they have been taking continuons steps towards using alternative and renewable energy sources.

On 9 April 2019, the German-Azerbaijani Chamber of Commerce (AHK Azerbaijan) released its "Market Analysis Azerbaijan 2019" publication in cooperation with Germany Trade & Invest (GTAI), which describes ways the South Caucasus' largest economy is working towards modernizing its conventional energy generation and electricity distribution sectors.
"Market Analysis Azerbaijan 2019" publication, which examined 15 sectors of the Azerbaijani economy including oil, gas, and other forms of energy, sees renewable energy as an area ripe for foreign investment. "Azerbaijan's energy sector is at a turning point: a law on renewable energy (RE) and an investor-friendly tariff policy are in sight and are soon likely to ensure a significant upswing in the RE industry," the document reads. "A number of projects represent potential opportunities for foreign suppliers of equipment and electricity industry know-how."

Azerbaijan teamed up with the Norwegian international consulting company DNV GL for drafting a support system and a legal framework to tap the potential of the alternative energy sector. The draft law will reportedly be completed by May. Foreign investors from China, Canada, Turkey, the United States and Norway, as well as Arab and EU countries, are supposedly eying opportunities in Azerbaijan with regards to renewable resources, especially wind power generation. According to the market analysis, wind power potential capacity accounts for over 15,000 megawatts (MW). Wind in Azerbaijan blows more than 250 out of 365 days per year and is able to generate over 2.4 billion kilowatt hours (kWh) of electricity per year.

But interest in green energy does not stop with wind. Solar power comes in at number two. Sunshine is present in Azerbaijan for anywhere between 2,400 – 3,200 hours per year.

The development of Azerbaijani's oil and gas sectors over the last two decades has turned the country into a reliable energy exporter, while at the same time creating potential for investing in and boosting the non-oil sector. The efforts of the government in Baku to attract private investment and cooperate with international energy companies and financial institutions, in order to develop the country's renewables capabilities and capacity, has been gaining momentum.

The Asian Development Bank (ADB) has allocated between $100,000 to $200,000 for a pilot project that will examine the feasibility, time and cost of a 300 kW floating solar power plant on Boyuk Shor Lake, located on the outskirts of Baku. The launch of the experimental facility is planned for March of 2021. The ADB is also investing in biogas plants, which will convert the waste produced during cotton, cereal and hazelnut cultivation into energy.

For its part, the World Bank is set to help the Ministry of Energy with funding small hydropower plants, with the help of the Canadian consulting company SNC-Lavalin Atkins that is analyzing the small hydropower market within the Caspian country. Azerbaijan currently has twelve big and seven small hydroelectric plants.

The European Bank for Reconstruction and Development is also interested in having its share in financial support campaign for renewable energy projects in Azerbaijan. In February, the French photovoltaic and wind power plant developer and operator Total Eren signed a memorandum of understanding with the energy ministry, which will see Total Eren lead projects that will generate 420 MW of power from wind, solar and bioenergy.

The six wind, 10 solar and six biomass power plants constructed between 2018 and 2020 in Azerbaijan are expected to have an installed capacity of about 420 MW, or an average annual electricity generation of about 1.2 billion kWh. The share of renewable energy in the domestic energy sector is expected to account for about 15 percent by 2020 according to the market analysis, and ultimately reach 40 percent in 2030.

Biodiversity research

Biodiversity is the variety of species, their genetic communities in which they occur. It is the duty of all the inhabitants of the planet to save the variety of biodiversity which forms the essence of the life.

Biodiversity Department established in the structure of the Ecology Department in 2017. Biodiversity Department serves to improve the management and monitoring system according to studying biodiversity, ecological systems and nature complexes, mitigation of impacts, biodiversity conservation, as well as, the protection of rare and endangered species in SOCAR's activity areas.

The Department of Biodiversity conducts monitoring in the following areas in accordance with procedural rules in the area of oil and gas operations:

- Soil and vegetation monitoring;
- Amphibians;
- Herpetofauna;
- Teriofauna;

Purpose of the research:

- Characterize the flora and fauna found in the study area.
- Assess the effects of operations on the area on flora and fauna.
- Comparing data from previous years with information to be obtained as a result of these researches.
- Determining the potential causes of any change in biodiversity, as well as taking into account both land-based operations and other reasons.

Hidrobiological research

The UN General Assembly Resolution 65/161 of 20 December 2010 proclaims the 2011-2020 period "The Decade of the United Nations for the implementation of the Strategic Plan for Biodiversity".
According to the Global Biodiversity Assessment (Global Biodiversity Assessment, UNEC, 1995),more than 30,000 animal and plants are at risk of extinction.
Throughout the entire geological history, to rate of extinction of mammals over the last 100 years has been 40 times higher than the maximum. The marine mammalian Caspian seal (phoca Caspica) is the only endemic species listed in the Red Book (Fifth National Report,2014)

During the monitoring on "Gunashly" oilfield our specialists discovered new species of cancroids – Kumkimi. Discovered species belong to "Volgocuma" class.

The discovery of Ecology Department's experts was patented.

During the monitoring, the food chain, species of mocroorganisms and their yearly changing is studied.

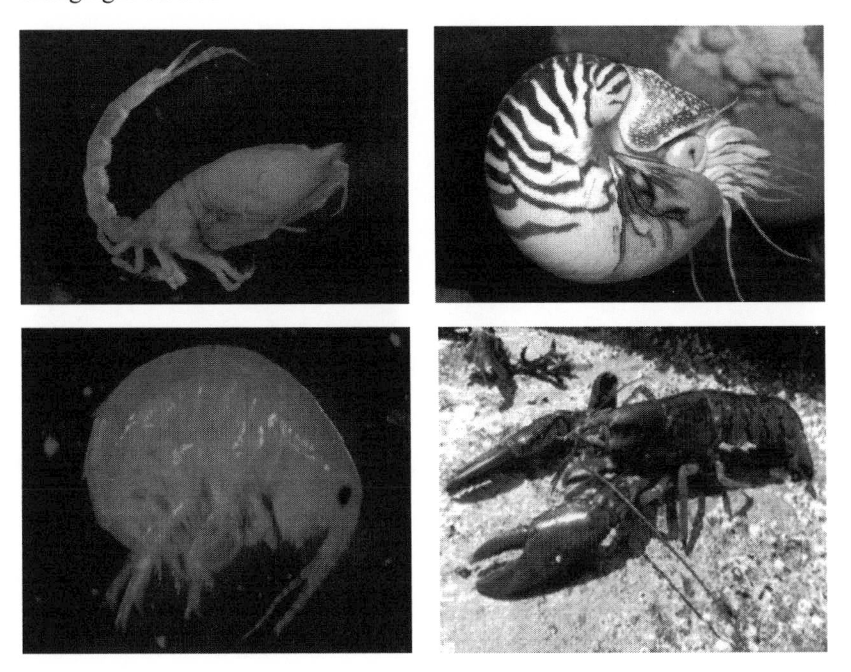

In connection with the study of biodiversity, the research of zooplankton, phytoplankton, periphyton, conducted during the ecological monitoring implemented in the Caspian Sea. Information acquired during etiologic, microbiological and ecotoxic researches can help study, assess and protect the composition of the food chain of the organisms, its dynamics of growth, inconstancy of the sea bottom and the dissemination area of fish of the water layer.

All sea plants and animals are included in the food chain of the Caspian Sea. During the ecological monitoring on the Caspian Sea, the dynamics of the food researched. chain is researched.

Samples collected in the course of ecological monitoring are analyzed in the laboratory,and reports are issued. SOCAR uses their Complex Research Laboratory for ecological studies, which is equipped with the latest analytical instruments and highly qualified professionals.

Hydrobiological studies- Identification of hydrocoles species and quantitative composition of each species of hydrocoles,representing the main food reserve for fish inhabition Caspian Sea or any other water body.These hydrocoles are followings: Zooplankton,Phytoplankton,Periphyton,Zoobenthos.

During offshore expedition hydrocoles are collected by means of special equipment (bathometer,plankton net,apparatus for collection of sea sediments), conserved in formalin and transported to the laboratory.In thelaboratory hydrocoles are studies with microscopes.

Ecotoxicological research

Ecotoxicological studies- Identification of impacts of pollutants to fish and hydrocoles in the sea.

There is created ecotoxicology area in Complex Research Laboratory that is, equipped with special devices for learning the influence of toxic substances drill sludges and oil to the creauters living in the sea and other waters. Total aroe laboratory is 49,6 m^2.It consists of 1 biotest room and 1 analysis of results room The camera that constructed for carrying out the ecotoxicological analysis has been equipped with constant lighting and constant temperature creating system. There is installed water tank for use and storage of sea water for carrying out ecotoxicological researches mechanical and ultraviolet water purification plant and Klimostat apparatus for carrying out biotests and cultivate of Clorella Vulgaris,Crustoceans.There is installed special aquarium, lighting and ventila- tion systems for long-term storage of creatures. Toxicity testing is carried out according to ISO DP 10253, ISO (1990)-TC 147/SG5WG2 standards at Ecotoxicology area of CRL.

Planting greenery activities

It is known that planting greens plays an important role in improving the environment. As a result of measures taken, 500 000 different trees and bushes were planted on SOCAR's areas of activity. Taking this into consideration, the Department of Ecology creates an "Ecological Park" on the area of 9.3 hectares on the territory of the Haji Zeynalabdin Tagiyev Oil and Gas Production Institute. The main purpose of creating the park is to create a modern form of public-ecological relations, to show that environmental propaganda is a social responsibility of citizens for the sake of environmental education and environmental protection. In the future, as part of environmental awareness activities, it is planned to organize student and student excursions to the park, environmental Olympiads, various quizzes and contests, as well as discussions on environmental issues. Solar batteries and wind generators will be used to irrigate the park, heat and heat the greenhouses. Thus, the use of alternative energy sources will save significant amounts of fuel and will also significantly reduce the amount of greenhouse gas emissions.

In general, the State Oil Company of the Republic of Azerbaijan attaches great importance to education and promotion.

The activity of the Ecology Department on greenery planting has two directions:

- Professional care and control of planting and landscape-design works;
- Consulting services for correct agro- technical care for the plants on SOCAR's production areas.

The Ministry of Ecology and Natural Resources has prepared a map of 650,000 tree-planting a day dedicated to the 650 th birth anniversary of the philosopher poet Imadaddin Nasimi. The map lists areas where trees will be planted in the regions of Azerbaijan, including the Nakhchivan Autonomous Republic.

A total of 650,000 trees in 37 species will be planted in 83 cities and districts as part of tree-planting campaing on the occasion of the 650 th birth anniversary of the great Azerbaijani poet Imadaddin Nasimi. This many trees will be planted is one day for the first time in Azerbaijan history.
Based on a map with activists of the New Azerbaijan Party's regional organizations has been started monitoring of tree- planting areas and control over the area will continue after the trees are planted. As part of the initiative, the Ministry of Ecology and Natural Resources plans to plant about 50 species of trees grown on vineyards. All seedlings, namely, Khan's plane tree, Eldar pine, cypress, acacia, ash tree, elm

tree, poplar, alder, willow, oleaster, wild pistachio, catalpa, olive tree, fig tree, peach tree, plum tree, apple tree, pomegranate tree, lime tree, plane tree, catalpa and other were grown in 17 hatcheries of the forestry conters and Gardening and landscape design of Azerbaijan OJSC. Most of which will be planted along the highways, which will serve to further the country's beauty. 50,000 trees are planned to be planted in the Shamakhi district and preparatory work is to be completed. Mainly during the fall tree season planting of honey giving trees designed.

They have also been discussed with the Beekeepers Association. Drip irrigation networks have already been laid on the tree plantations in Shamakhi and Sabirabad. The laying of that network means a high percentage of trees planted in that area. Also the previously contaminated area of the iodine-bromine plant in Surakhani district of Baku has been cleared and planted trees are also planned there.

This campaing will help increase the number of trees in the country and reduce the impact of climate change. On the initiative of First Vice President of the Republic of Azerbaijan Mehriban Aliyeva 650,000 trees will be planted in the country on the occasion of the 650th anniversary of the famous poet Imadaddin Nasimi. The initiative, covering all districts of the country, was launched on December 6 ,2019 in Shamakhi , where the famous poet was born. President of the Republic of Azerbaijan Ilham Aliyev and First Lady Mehriban Aliyeva have attended a tree-planting campaign in Shamakhi district and they planted trees.President Ilham Aliyev and First Lady Mehriban Aliyeva then posed for photographs together with the young people attending the tree-planting campaign.
At the some time, this is a valuable contribution aimed at uniting people around the initiative, showing the public solidarity in improving the environment.

Investigation and introduction of new plants and seeds of the East Asian flora in the Absheron peninsula.

The steady growth of oil and gas production in the world and the dynamic development of industries have become the main cause of environmental degradation on our planet, with negative impacts on the environment. In oil-rich Azerbaijan, for many years, large land plots have become contaminated by oil spills. The disturbance of the ecological balance in the Absheron peninsula, where the mining industry is developing, is more vivid.

Therefore, the regulation of the environmental situation in the oil and gas production areas of the peninsula has become an important issue of the day.Based on the foregoing, the introduction of new ornamental plants to Azerbaijan is one of the most important issues facing researchers in the Absheron environment to study their biological characteristics depending on climatic factors, identify promising species, and identify opportunities for their use in other industries.In this regard ,the topic is relevant in the Absheron conditions in terms of investigation the eco-biological features of same ornamental trees and shrubs of the East Asian flora,identifiying introductory opportunities and decorative features, and evaluating the prospects for its use in other industries.
In view of the foregoing ,the aim of the research was to study the bioecological features of new species of East Asian flora in the Absheron conditions and to apply the most promising species in greening. In this regard ,the selection of trees and shrubs according to local conditions,the determination of their taxonomic composition, their intrusion and mobilization, and their relationship to the environmental and climatic factors in the Absheron peninsula is of great scientific and practical importance.

The aim of the study was to carry out a number of scientific and practical activities aimed at the study and implementation of bioecological features of the newly introduced East Asian flora taxas.
At present, more than 200 species of the East Asian flora trees and shrubs have been introduced in the Central Botanical Garden of the Azerbaijan National Academy of Sciences.However,the most of these species has not yet been studied in the new conditions their eco-biological prosperties.
For the first time, 14 new species plant seeds (Hydrangea bretschneideri, Abies sachalinensis, Picea jezoensis, Buddleja japonica, Aralia chinensis, Picea gemmata, Maackia amurensi, Tetradium daniellii, Platycarya strobilacea, Abies koreana, Acer

pseudosieboldianum, Prunus avium, Padus maackii, Symplocos paniculata) brought from our side, cultivated in the Central Botanical Garden of the Azerbaijan National Academy of Sciences. In addition, 6 new species (Padus maackii , Thujopsis dolbrata., Microbiota decussata, Clematis fusca , Clerodendrum trichotomum , Photinia davidiana) brought from our side were planted in closed space. The introduced plants and seeds were brought from the General Botanical Garden of the Russian Academy of Sciences. Many researchers have published a number of works on the reproduction of trees and shrubs. In different months of the year, the seeds were sown in an open and indoor area to replenish the seeds we planted. Seeds of a number of trees and shrubs begin to germinate in the first year, and in some species in the second year, depending on the silence of the seeds. With this in mind, the seeds were stratified before sowing and sown in January-April and September-December. The results of the experiments showed that the seeds of most of the plant species studied do not get high germination rates when they are sown in March-April.Table1. Reproduction of some trees and shrubs of East Asian flora in ex situ conditions. From the experiments it was clear that the seeds of most of the plants studied should be stratified.

№	Species	*ex situ*							
		Percentage of monthly sowing and germination							
		I	II	III	IV	IX	X	XI	XII
1.	*Abies koreana*	13	22	26	31	0	0	3	9
2.	*Abies sachalinensis*	11	16	23	33	0	0	5	7
3.	*Acer pseudosieboldianum*	24	23	18	17	31	29	28	31
4.	*Aralia chinensis*	16	21	24	6	7	9	13	18
5.	*Buddleja japonica*	62	59	46	67	23	25	18	29
6.	*Hydrangea bretschneideri*	53	54	42	53	8	11	16	16
7.	*Maackia amurensis*	31	33	38	54	14	26	31	26
8.	*Padus maackii*	72	61	50	41	74	69	80	79
9.	*Picea jezoensis*	11	13	9	0	0	7	13	16
10.	*Picea gemmata*	16	13	8	2	0	5	18	17
11.	*Platycarya strobilacea*	16	63	56	37	63	59	69	73
12.	*Prunus avium*	66	68	65	64	78	83	88	87
13.	*Symplocos paniculata*	19	23	39	31	3	11	13	13
14.	*Tetradium daniellii*	0	1	1	3	0	0	0	1

Otherwise, the seeds that have not been stratified are almost certainly not germinated when sown in March-April. The analysis showed that Tetradium daniellii was the lowest germination percentage in the study plants. The analysis of the literature data shows that generally, the survival of the seeds of the bi species is very low. That is one of the 4 seeds that has been sown has the chance of germination. The results of experimental work have shown that different results have been obtained depending on the individual biological characteristics of each species studied. The results obtained for some species are not satisfactory(Acer pseudosieboldianum, Tetradium daniellii, picea jezoensis,picea gemmata). Generally, it is possible to conclude that reproduction of species studied under ex situ conditions can be carried out and a large number of seeds can be extracted from them.

Phenological observations by us have shown that the species of plants that have been introduced into the fungi continue to grow and develop normally under new conditions. Experimental work in this direction continues.

Ecological enlightenment and propaganda

One of the activities of the Ecology Department is ecological enlightenment. Environmental Education, Enlightenment and Environmental Literacy – are awareness activity system directed to the formation of ecological culture that are focused changing people's thinking and development of new behavioral stereotypes. Additionally, it will be helped changing their lifestyle and ecologically forming a world view. SOCAR pays more attention to environmental enlightenment accordance to "The legislation on environmental education and enlightenment of population" signed by National Leader Heydar Aliyev. Portentous dates the environmental on protection are broadly celebrated at SOCAR's enterprises environmental trainings are organized for the employees. Meanwhile, cooperation possibilities with international environmental organizations, as well as governmental and non-governmental organizations are increasing widely.Ecology Department signed a cooperation agreement in the field of environmental awareness between Baku City Education Department on 30th January with accordance to involve a wider audience in environmental enlightenment and propaganda; to organize systematic, effective and events more sustainable. "Ecological months", “Green school" projects and other projects are being implemented with Ecology Department in the direction of "Ecology and Health".

1500 trees were distributed to 40 schools of Baku city by Ecology Department within the framework of the "Green school" project with accordance to the agreement. "Alternative and renewable energy use", "Healthy lifestyle", "Energy Efficiency", “Biodiversity of Azerbaijan" and on others topics presentations were held in schools of Baku city by Ecology some Department's expert.The excursions were organized to "Eco-park Environmental Research Center" Public Union, "Fire Temple of Baku" State Historical-Architectural Reserve and "Qala Archaeological and Ethnographic Museum Complex" for schoolchildren.

Literature

1.ALLEN,K.R.1955. The growth of accuracy in ecology. Proc. New Zeal.Ecol.Soc. 1:1-7.
2.ASHBY, E. 1948. Statistical ecology. II. A reassessment. Bot. Rev. 14:222-234.
3.ATTIWILL,P. 1981. Energy, nutrient flow and biomass. Proc. Austral. Forest Nutr. Workshop 1:131-144. CSIRO, Melbourne.
4.BAZZAZ, F.A 1979. The physiological ecology of plant succession. Ann. Rev. Ecol. Syst. 10:351-371.
5.BERRYMAN, A. A 1981. Population system: A General Introduction. Plenum, New York.
6.BJORKMAN, O.1973 .Comparative studies on photosynthesis in higher plants. In Photophysiology, vol. 8, ed. A. C. Giese, pp. 1-63. Academic Press, New York.
7.BILLINGS, W. D 1952. The environmental complex in relation to plant growth and distribution. Quart. Rev . Biol. 27:251-265
8.BJORKMAN, O. 1975. Inaugural address. In Environmental and Biological Control of Photosynthesis, ed. R. Marcell ,pp. 1-16. Dr.W. Junk Publishers, The Hague.
9.BOYKO, H.1947. On the role of plants as quantitative climate indicators and the geoecological law of distribution . J . Ecol. 35:138-157.
10.BRAUN-BLANQUET, J.1932. Plant Sociology, transl. G.D. Fuller and H.S. Conard. McGraw-Hill,New York.
11.BRIAND, F. 1983. Environmental control of food web structure. Ecology 64:253-263.
12.BRODY, S.1945. Bioenergetics and Growth.Van Nostrand Reinhold, New York.
13.BROWER, L.P.1969. Ecological chemistry .Sci. Amer. 220(2):22-29.
14.BURKE, M. J., GUSTA, H. A. QUAMME, C. J.WEISER, and P.H.LI.1976. Freezing and injure in plant. Ann. Rev. Plant Physiol. 27:507-528.
15.CAIN, S. A.1944. Foundations of Plant Geography. Harper & Row, New York.
16.CAVERS, P. B., and J. L. HARPER. 1967. Studies in the dynamics of plant populations. I .The fate of seed and transplants introduced into various habitat. J. Ecol 55:59-71.
17.CAUGHLEY ,G., and J.H.LAWTON 1981. Plant-herbivore systems. In Theoretical Ecology ,ed. R.M.May,pp.132-166. Blackwell, Oxford, England.
18.CHAPMAN, R. N. 1928. The quantitative analysis of environmental factors. Ecology 9:111-122.
19.CLEMENTS, F. E. 1916. Plant Succession: An Analysis of the Development of Vegetation. Carnegie Inst. Publ. No .242, Washington, D . C.
20.CLEMENTS, F. E. 1936. Nature and Structure of the climax. J. Ecol. 24:252-284.

21.CLEMENTS. F. E.1949. Dynamics of Vegetation. Macmillian (Hafner Press), New York.
22.COLINVAUX, P. 1973. Introduction to Ecology. Wiley, New York.
23.CONNELL, J. H., and E. ORIAS.1964. The ecological regulation of species diversity. Amer. Nat. 98:399-414.
24.CONWAY, G. 1981. Man versus pests. In Theoretical Ecology, ed. R.M.May,pp. 356-386. Blackwell, Oxford, England.
25.COTTAM, G., and R. P. MC INTOSH. 1966. Vegetation continuum. Science 152:546-547.
26.DOUBENMIRE, R. F. 1974. Plants and Environment, 3d ed. Wiley, New York.
27.DE BACH, P., ed. 1964. Biological Control of Insect Pest and Weeds. Chapman & Hall, London.
28. DE BACH, P. 1974. Biological Control by Nature Enemies. Cambridge University Press, London.
29.DEVLIN, R. M. 1969. Plant Phsiology. Van Nostrand Reinhold, New York.
30.ISKENDER E. O., ZEYNALOV Y., OZASLAN M., et all. Investigation and introduction of same rare and threatened plants from Turkey // J. Biotechnology & Biotechnological Equipment, 20/2006/3 , p . 60-68.
31.IUCN 1994. IUCN Red List Categories. IUCN Species Survival Comission. IUCN, Gland, Switzerland and Cembridge, UK, 1994, 21pp.
32.IUCN 2001. IUCN Red List Categories and Criteria: Version 3.1 IUCN Species Survival Comission. IUCN, Gland Switzerland and Cambridge, 2001,ii+30pp.
33.IUCN 2003. Guidelines for Re-introduction, Prepared by the IUCN/SSC Re-introduction Specialist Group. IUCN, Gland Switzerland and Cambridge, UK, 1998, 11p.
34.IUCN 2003 . Guidelines for Application of IUCN Criteria at Regional Levels. Version 3.0 IUCN Species Survival Commission. IUCN, Gland, Switzerland and Cambridge, 2003, UK.ii+26pp.
35. EGERTON, F .N., III. 1973. Changing concepts of the balance of nature. Quart. Rev. Biol. 48:322-350.
36 .ELTON, C. 1958. The ecology of Invasion by Animal and Plants. Methuen, London.
37.EHRLICH, P. R., and P. H. RAVEN. 1964. Butterflies and Plants: A study in coevolution. Evolution 18:586-608.
38.ETHERINGTON, J. R. 1975. Environment and Plant Ecology . Wiley, New York.

39 .EYRE, S . R. 1963. Vegetation and Soil: A World Picture. Aldine, Chicago.
40.FORD, E. B. 1975 Ecological Genetics, 4 th ed. Chapman & Hall, London.
41.FRANKLIN, I R. 1980. Evolutionary change in small populations. In: Soule, M. Wilcox, B. (ed), Conservation Biology: An Evolutionary Ecological Perspective, Sinauer Associates, Sunderland, MA, pp. 135-149.
42.GOLLEY, F. B. 1961. Energy values of ecological materials. Ecology 42:581-584.
43.GOOD, R. 1964. The Geography of the Flowering Plants. Longmans, London.
44.GREIG-SMITH, P. 1964. Quantitative Plant Ecology, 2d ed . Butterworth, London.
45.GREiG- SMITH, P. 1979. Pattern in vegetation. J. Ecol. 67:755-779.
46.HARPER, J. L. 1977. Population Biology of Plants. Academic Press, New York.
47.HARPER, J. L., P. H. LOVELL, and K. G. MOORE. 1970. The shapes and size of seeds. Ann. Rev. Ecol. Syst 1:327-356.
48.HARPER, J. L., and J. WHITE. 1974. The demography of plants. Ann. Rev. Ecol. Syst. 5:419-463.
49.HARVEY, P. H., R. K. COLWELL, J. W. SILVERTOWN, and R. M. May. 1983. Null models in ecology. Ann. Rev. Ecol. Syst. 14:189-211.
50.HEINRICH, B., and P .H. RAVEN. 1972. Energetics and pollination ecology. Science 176:597-602.
51.HESSE, R., W. C. ALLEE, and K. P. SCHMIDT. 1951 . Ecological Animal Geography , 2d ed. Wiley, New York.
52. HOWE, H. F., and J. SMALLWOOD. 1982. Ecology of seed dispersal . Ann. Rev. Ecol. Syst. 13:201-228
53.HUTCHINSON, G. E. 1978. An Introduction to Population Ecology. Yale University Press, New Haven, Coon.
54.KERSHAW, K. A. 1973. Quantitative and Dynamic Ecology. Edward Arnold, London.
55. KINNE, O., ed 1970. Temperature. In Marine Ecology ,vol 1: Enviromental Factors, pt. 1, chap .3. Wiley – Interscience, New York.
56.KLEIBER, M. 1961. The Fire of Life. Wiley, New York.
57.LARCHER, W. 1980. Physiological Plant Ecology, 2d ed. Springer- Verlag, Berlin.
58.LEVIN, D. A. 1976a. Alkaloid-bearing plants: An ecogeographic perspective. Amer. Nat. 110:261-284.
59.LUKE J. H, STATTON B. Conservation of Small Populations: Effective

Population Sizes. Inbeerding, and the 50/500 Rule.
60.LOTKA, A. J. 1925. Elements of Physical Biology. (Reprinted in 1956 by Dover Publications, New York.)
61.MACAN, T. T. 1974. Freshwater Ecology, 2d ed. Longmans, London.
62.MAJOR, J. 1958 Plant ecology as a branch of botany. Ecology 39:352-363.
63.MAJOR, J. 1963 A climatic index to vascular plant activity. Ecology 44:485-498.
64.MARK L. SHAFFER. Minimum population sizes for species conservation. BioScience, 1981.
65.MARGALEF, R. 1958. Information theory in ecology. Gen. Syst. 3:36-71.
66. MARGAlEF, R. 1968. Perspectives in Ecological Theory. University of Chicago Press, Chicago. 67.MILLER, R S. 1957. Observation on the status of ecology. Ecology 38:353-354.
68.MUELLER-DOMBOIS, D., and H. ELLENBERG. 1974. Aims and Methods of Vegetation Ecology. Wiley, New York.
69.MURPHY, G. I. 1968. Pattern in life history and the environment. Amer. Nat. 102:391-403
70.ODUM, E. P. 1963. Ecology. Holt, Rinehart and Winston, New York.
71.PARKER, J. 1963. Cold resistance in woody plants. Bot. Rev. 29:123-201.
72.PARKER, J. 1969. Further studies of drought resistance in woody plants. Bot. Rev. 35:317-371.
73. PEARL, R. 1928. The Rate of Living. Knopf, New York.
74. PHILLIPS, E, A. 1959. Methods of Vegetation Study. Holt, Rinehart and Winston, New York.
75.PIANKA, E. R. 1974. Evolutionary Ecology. Harper & Row, New York.
76.PITTOCK, A .B., and M. J. SALINGER 1982. Towards regional scenarios for a CO2- warmed earth. Climatic Change 4:23-40.
77.POOLE, R .W., and B. J. RATHCKE. 1979. Regularity, randomness, and aggregation in flowering phenologies . Science 203:470-471.
78. POORE, M. E. D. 1956. The use of phytosociological methods in ecological investigations. IV. General discussion of phytosociological problems. J. Ecol. 44:28-50.
79. REGAL, P. J. 1982. Pollination by wind and animals: Ecology of geographic patterns. Ann. Rev. Ecol. Syst 13:497-524.
80.RIDLEY, H. N. 1930. The Dispersal of Plant Throughout the World. L. Reeve, Ashford, Kent, England.

81.SANDERS, H. L. 1968. Marine benthic diversity: A comparative study. Amer. Nat. 102:243-282. 82.SEAN S, CLINTON N. J, LUCAS N. J, DAVID L. A. G, WILLIAM F. L / Remaining natural vegetation in the global biodiversity hotspots, 2014.
83.TREWARTHA, G. T. 1954. An Introduction to Climate ,3d ed. McGraw- Hill, New York.
84.TURESSON, G. 1922 . The species and the varity as ecological units. Hereditas 3:100-113.
85. TURESSON. G. 1925. The plant species in relation to habitat and climate. Hereditas 6:147-236.
86. TURESSON. G. 1930. The selective effect of climate upon the plant species. Hereditas 14:99-152.
87.WILLIAMS, C. B. 1964. Patterns in the Balance of Nature and Related Problems in Quantitative Ecology. Academic Press, New York .
88.ZELITCH, I 1971. Photosynthesis, Photorespiration, and Plant Productivity. Academic Press, New York.

Internet resources:

1. http://www.socar.az
2. http://www.trend.az
3. http://www.caspiannews.com
4. http://www.ebrd.com
5. http://www.az.undp.org
6. http://www.unece.org
7. http://www.report.az
8. http://www.cms.int
9. http://www.cms.int/en/node/3916
10. http://cbd.int/doc/world/az/az-nr-05-en.pdf
11. http://whc.unesco.org
12. http://www.unenvironment.org
13. http://www.unfccc.int
14. http://www.cbd.int/ibd/2005/default.shtml
15. http://www.eea.europa.eu/ru/publications/environmental.../ru_11_0.pdf
16. http://www.chd.int/doc/world/az/az-nr-05-en.pdf
17. http://www:eea.europa.eu
18. http://www.eco.gov.az
19. http://www.species360.org
20. http://www.cbd.int/information/parties.shtml
21. http://www.cbd.int/doc/legal/cbd-en.pdf
22. http://www.cbd.int/idb/2011
23. http://www.theseasproject.weebly.com
24. http://www.cbd.int/idb/2012
25. http://www.cbd.int/undb/home/undb-strategy-ru.pdf
26. http://www.cbd.int/idb/2013
27. http://www.iucnredlist.org
28. http//www.cbd.int/idb/2007/default.shtml
29. http//www.conventions.ceo.int/Treaty/en/Threaties/Html/104.htm
30. http://www.cbd.int/idb/2014
31. http://www.iucn.org
32. http://www.cbd.int/doc/strategic-plan/2011-2020/Aichi-Targets-en.pdf
33. http://www.cbsg.org
34. http://www.cbd.int/doc/bioday/2007/ibd-2007-booklet-01-ru.pdf
35. http://www.conservation.org

36.http://www.enccd.int
37.http://www.international-climate-initative.com
38.http://www.bgci.org
39.http://www.iwc.int /convention
40.http://www.diversitas-international.org
41.http://www.cites.org
42.http://www.cites.org/eng/disc/text.php
43.http://www.osce.org
44.http://www.biodat.ru/doc/lib/degkin2.htm
45.http://www.esri.com
46.http://www.un.org/documents
47.http://www.fao.org
48.http://www.usembassy.gov
49.http://www.uia.org
50.http://www.ramsar.org

Baghirova Shafaq – I was born on April 3, 1990 in Lenkaran district of Azerbaijan. I entered the Biology faculty of Baku State University in 2008 and I graduated the full course in biology from the BSU in 2012. In 2013 I entered the faculty of Method and Methodology teaching in Biology at the ADPU as a master. In 2015 I graduated there with a Master science specialty degree. Since 2019, PhD student in ecology at the BSU. I am studying for a PhD degree in the subject "Eco-biological features and perspective of some woody plants of the East Asia flora in the Absheron conditions".

FSC
www.fsc.org
MIX
Papier aus verantwortungsvollen Quellen
Paper from responsible sources
FSC® C141904

Druck:
Customized Business Services GmbH
im Auftrag der
KNV Zeitfracht GmbH
Ein Unternehmen der Zeitfracht - Gruppe
Ferdinand-Jühlke-Str. 7
99095 Erfurt